The National Environmental Performance Partnership System
A Review of Implementation Practices

United States
Environmental Protection
Agency

Office of the Administrator
EPA 140-R-13-001
May 2013

Table of Contents

I. INTRODUCTION

EPA and the states have more than fifteen years of experience implementing the National Environmental Performance Partnership System (NEPPS) to organize the federal-state relationship in terms of setting priorities, deploying resources, and measuring progress. The EPA-state partnership to protect human health and the environment has matured and improved substantially during that period.

This review assesses how and to what extent NEPPS implementation has helped to realize the goals for strengthening the EPA-state partnership, articulated below, characterizing both the progress NEPPS has made since 1995 and the challenges the program faces going forward. The review concludes with recommendations on addressing obstacles facing NEPPS while building upon its past successes. The findings and recommendations in this report reflect EPA's perspective on and views of NEPPS; it does not represent the states' current views or thinking on this topic. EPA recognizes that it is essential to work with the states to prioritize the recommendations for implementation, with the goal of further improving the EPA-state shared governance framework and promoting a greater mutual consensus on priorities, planning and accountability.

This review is supported by input from various sources including reports, briefings, surveys, and other documents:

- Government Accountability Office (GAO) and Office of the Inspector General (OIG) reports on NEPPS-related topics
- NEPPS Program Implementation Data
- Office of Congressional and Intergovernmental Affairs (OCIR) issue papers and briefing materials
- NEPPS National Guidance (FY 2005-2013)
- Interviews with the Deputy Regional Administrators (DRAs) on NEPPS and the EPA-State Partnership, Fall 2011 and Winter 2012
- EPA-State Strategic Planning Pilots (March 2005 and December 2009)
- *White Paper on Meeting the Challenges of Environmental Protection Together: Building Strong EPA-State Partnerships*, EPA, October 2010

The original NEPPS Agreement, the *Joint Commitment to Reform Oversight and Create a National Environmental Performance Partnership System*, (hereinafter 1995 NEPPS Agreement) outlined criteria by which the program is to be periodically evaluated: effectiveness, public credibility, fiscal soundness, and program accountability. This analysis supports three goals articulated in the *FY 2011-2015 EPA Strategic Plan Cross-cutting Fundamental Strategy for Strengthening State, Tribal and International Partnerships*.

1. Improve implementation and consistent delivery of national environmental programs through closer consultation and transparency.

2. Work with states to seek efficient use of resources through worksharing, joint planning using data analysis and targeting to address priorities, and other approaches.

3. Play a stronger management role to facilitate the exchange of data with states to improve program effectiveness and efficiency.

II. BACKGROUND

A. The EPA-State Relationship and NEPPS

The EPA-state relationship has long been complex, due in part to the intricate division of roles and responsibilities under the various federal environmental statutes. Primary responsibility for 37 programs under 8 statutes can be delegated to states. These laws assign to EPA responsibility to promulgate regulations, develop policy and guidance, oversee permitting and enforcement, collect data from the states, measure performance, and report to Congress. States are often charged with carrying out most of the implementation work, including issuing permits, providing compliance assistance, inspecting regulated facilities, and initiating enforcement actions.

Prior to the creation of EPA in 1970, states provided the majority of environmental management controls—permits, discharge standards, as well as public health and natural resource regulations. Subsequent to the establishment of EPA, the federal government became an integral partner with states and localities in managing, promoting and regulating environmental protection activities. By the 1990s, states had increased their environmental management capacities and were maintaining core program activities. Growing interest in performance-based management and increasing recognition that remaining environmental problems required innovative solutions set the stage for the 1993 Task Force on Enhancing State Capacity.

In its report, the Task Force set forth a number of recommendations including establishing: a new framework and policy for EPA-state relations; a joint process for strategic planning and integration of priorities; and a mechanism to institutionalize state capacity. The Task Force also recommended further strengthening state management capacity and infrastructure; streamlining the grants assistance process; pursuing alternative financing mechanisms; and encouraging legislative action to accomplish various ends (e.g., proposing language for legislative initiatives to make state-capacity building a primary mission for EPA; proposing legislative changes to the Administrative Procedures Act (APA)[1] and the Federal Advisory Committee Act (FACA)[2]; seeking amendments during reauthorization of EPA's statutes to clarify the roles and responsibilities of the states and EPA.

The Task Force gave rise to the State/EPA Capacity Steering Committee which was charged with implementing the Task Force's recommendations. The Committee produced the 1995 NEPPS Agreement which outlined the following guiding principles:

---

[1] The Task Force recommended: "Articulate the current limits and opportunities under the APA for including states in the rule-making process, and propose specific legislative changes to the APA that would address EPA and state needs."

[2] The Task Force recommended: "Offer guidelines on how EPA can currently work with the states under the FACA. Propose specific changes to FACA that would recognize the right of states, as delegated managers of EPA programs, to be consulted on matters of policy and management of national environmental programs without the need to charter formal advisory committees."

1. Continuous environmental improvements are desirable and achievable throughout the country.

2. A core level of environmental protection should be maintained for all citizens.

3. National environmental progress should be reported using indicators that are reflective of environmental conditions, trends, and results.

4. Joint EPA/state planning should be based on environmental goals that are adaptable to local conditions while respecting the need for a "level playing field" across the country.

5. EPA/state activity plans and commitments should allocate federal and state resources to the highest priority problems across all media, and should seek pollution-prevention approaches before management, treatment, disposal, and cleanup.

6. A differential approach to oversight should provide an incentive for state programs to perform well, rewarding strong state programs and freeing up federal resources to address problems where state programs need assistance.

As states have improved their management capacity, the number of eligible programs delegated to states has increased. From 1992 to 2007 eligible program delegation to states has increased from 40 to 96%. Now states conduct over 90% of core program activities, including permitting, compliance enforcement, data entry, first response, and developing non-regulatory tools. On average, 25% of state environmental budgets are funded by federal grants; however, states have assumed an increasing share of implementation costs.[3]

The EPA-state relationship has evolved considerably since the inception of NEPPS. While states once had minimum influence in goal and priority setting, and state commitments were determined through cumbersome memoranda of agreement (MOAs)[4] between EPA headquarters and the regions, states now have opportunities to influence priorities, strategies, and annual performance commitments. They are more involved with budget development and most negotiate Performance Partnership Agreements (PPAs) and Performance Partnership Grants (PPGs), the primary tools to establish priorities and deploy resources. PPGs allow states to combine categorical grants for greater spending flexibility on state priorities. PPAs are negotiated strategic plans that articulate joint goals and priorities, key activities and roles and responsibilities.[5]

Beginning with the original 1995 NEPPS Agreement, and affirmed over time in various vision statements, operating principles and Environmental Council of States (ECOS) resolutions, states have consistently expressed their support for NEPPS. For example, in March 2011, ECOS

---

[3] Data source: Environmental Council of States. See: http://www.ecos.org/content/search?search_query=program+delegation.

[4] MOAs were replaced by EPA's Annual Commitment System (ACS) in 2004.

[5] For more information about PPAs and PPGs, see: http://www.epa.gov/ocir/nepps/index.htm.

passed a resolution that affirmed "NEPPS overall as an important and positive step in the right direction," and its role in providing benefits for many states through the use of PPAs and PPGs, "including increased administrative flexibility, the ability to direct grant resources to cross-cutting and multi-media projects under Part 35 grant rules, and increased support for innovative projects."[6]

Further, the resolution reaffirmed support for the principles embodied in the original 1995 NEPPS Agreement and for the gains realized through NEPPS, including those achieved through the use of PPAs and PPGs, and recognized that NEPPS has served to assist EPA in meeting the goals of the Government Performance Results Act (GPRA). It also acknowledged that there are opportunities to realize even greater nationwide benefits from the existing NEPPS model, and that doing so is more important than ever given current, and most likely future, federal and state budget challenges.

Tracking the progress and performance of states using PPAs and PPGs has been an important feature of NEPPS since its inception in 1995. Figure 1 illustrates a slow but steady increase in the state environmental agency participation in NEPPS (via the use of PPAs and PPGs) for fiscal years 1997-2012. In FY 2012, 33 states and territories were using PPAs and 42 were using PPGs.

During the same time frame, participation by state agriculture agencies in NEPPS (Figure 2) shows a decline in PPA use (from a high of 8 in FY 2000 to a low of 2 in FY 2008), but an increase in PPG use (from a low of 16 in FY 1997 to a high of 29 in FY 2003 and 2004). Many state agriculture agencies are responsible for implementing three EPA pesticides grants: Pesticides Program Implementation (FIFRA 23(a)(1)); Pesticides Cooperative Enforcement (FIFRA 23(a)(1)); and Pesticide Applicator Certification and Training (FIFRA 23(a)(2)). These state agriculture agencies combined their pesticide grants into PPGs by taking advantage of the administrative benefits and flexibility afforded by this tool.[7]

---

[6] ECOS Resolution Number 8-10, Revised March 30, 2011: *Continued State Commitment to NEPPS and Strengthening the State-EPA Partnership.*

[7] In addition, eight state environmental agencies responsible for implementing pesticide grants manage them in PPGs.

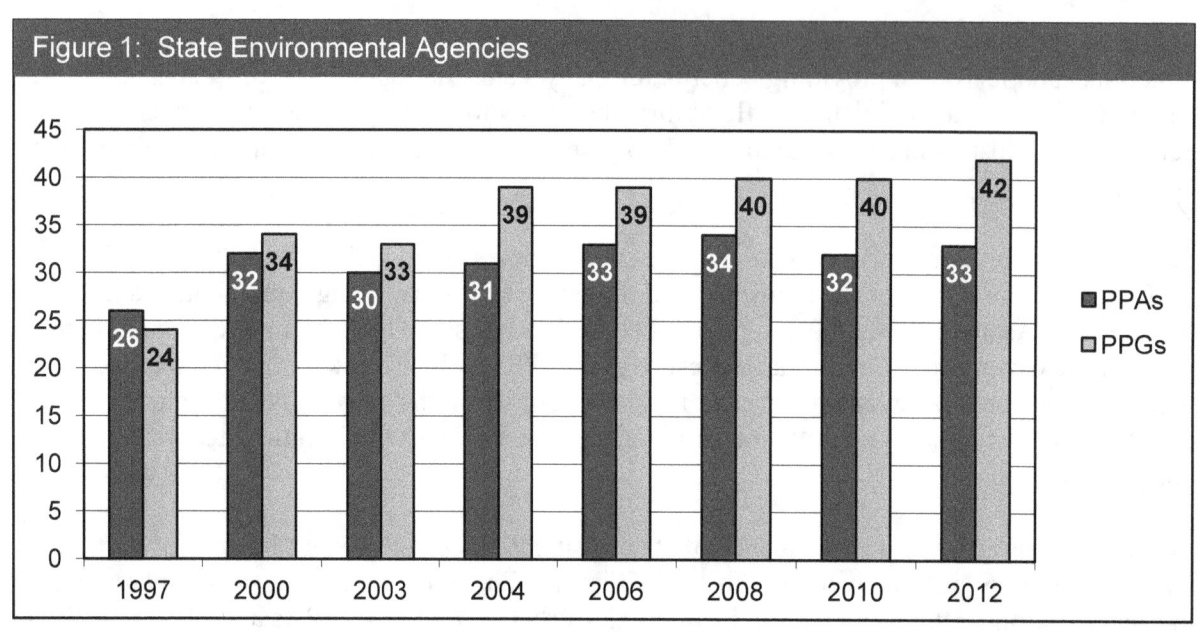

Figure 1: State Environmental Agencies

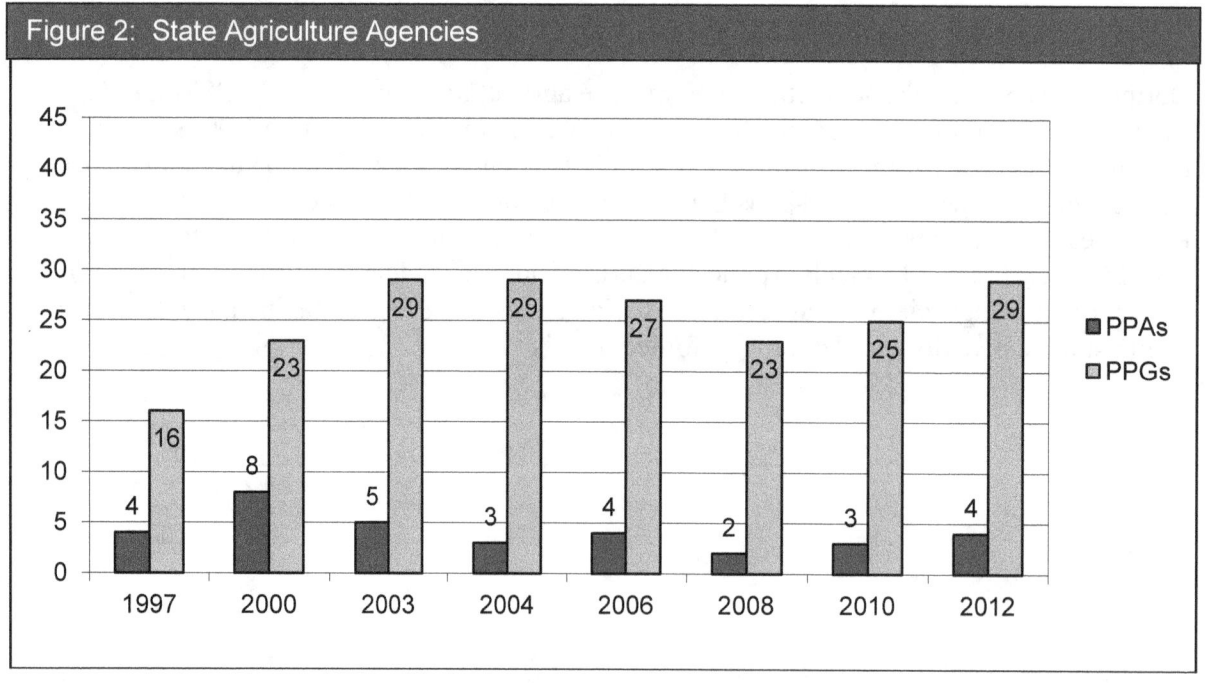

Figure 2: State Agriculture Agencies

A review of PPAs used by the states in EPA's 10 regions in FY 2012 reveals that there are three prevailing types with the following characteristics:

- Type A: comprehensive, strategic and operational and covers all major program areas; often linked to grant resources, provides for administrative and/or programmatic flexibility; and serves as the PPG and/or categorical grant workplan.

- Type B: broad in scope, enumerates the state's strategic priorities, and may be linked to grant resources; may serve as the PPG workplan; may include a description of environmental conditions and strategies for addressing priorities, roles and responsibilities; and may include compliance and enforcement provisions.

- Type C: high-level strategic agreement; generally not linked to grant resources and does not serve as a PPG workplan; includes the state's strategic priorities and may include compliance and enforcement provisions.

In recent years, slightly over half of all the PPAs have been Type A (comprehensive and operational), while the balance typically comprised Types B (broad) and C (high-level and strategic) PPAs. Figure 3 shows the distribution of PPA types by region.[8]

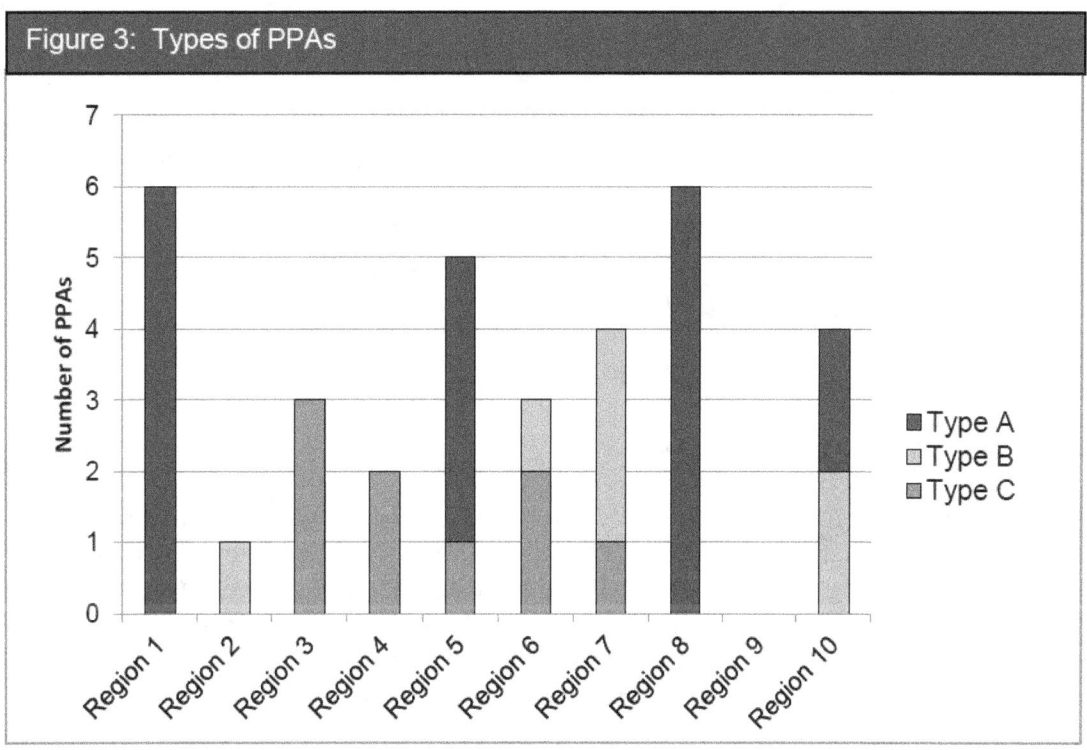

Figure 3: Types of PPAs

Figure 4 illustrates the rate of growth in the amount of eligible State and Tribal Assistance Grant (STAG) funds that are managed in PPGs. In FY 2012, 38 % of eligible funds were included in PPGs.

---

[8] Total number of states/territories in each region: R1(6 states); R2 (2 states/2 territories); R3 (6 states); R4 (8 states); R5 (6 states); R6 (5 states); R7 (4 states); R8 (6 states); R9 (4 states/2 territories); R10 (4 states).

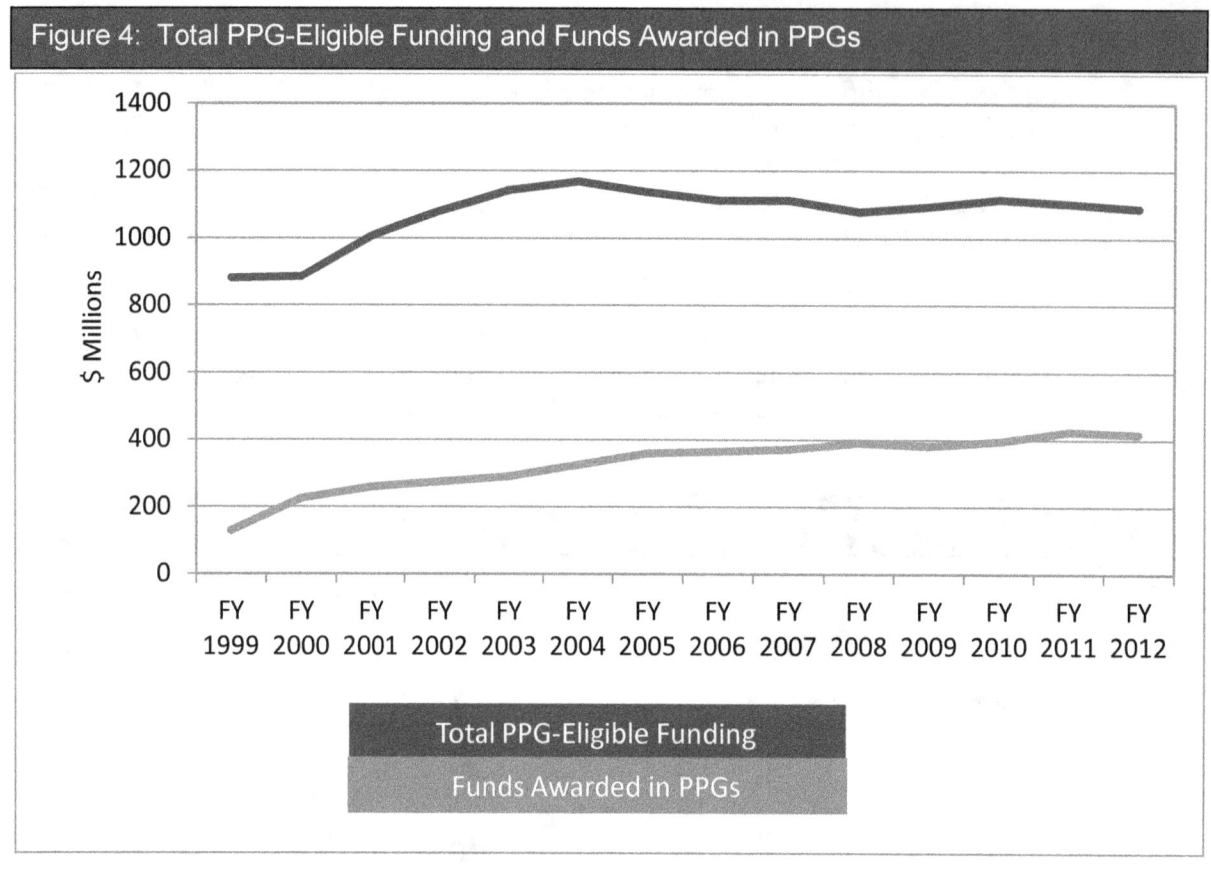

Figure 4: Total PPG-Eligible Funding and Funds Awarded in PPGs

B. External Reports Evaluating NEPPS

During the past decade, the OIG, GAO, the National Academy of Public Administration (NAPA), ECOS and others have evaluated various aspects of NEPPS. These early reports revealed certain recurring themes and many were quite critical. For example, GAO noted that NEPPS had not been well integrated into EPA; that states indicated their investment in NEPPS outweighed the benefits received; OIG concluded that EPA leadership did not provide clear direction and set expectations with regard to NEPPS; that a lack of training of EPA staff had hindered effective implementation of NEPPS; ECOS noted that the PPA/PPG experience yielded mixed results; NAPA stated that top EPA officials gave inconsistent support to NEPPS.

The reports concluded with recommendations for improvements and changes, most of which EPA has since adopted or addressed. The recommendations cited in the early reports from GAO and OIG included concepts and ideas that were later developed and ultimately codified in the PPG regulations at 40 CFR Part 35 in 2001. A 2007 GAO report noted that EPA has made substantial progress in improving priority setting and enforcement planning with states through its system for setting national enforcement priorities and NEPPS, thereby fostering a more cooperative relationship. Also, EPA and states have made progress in using NEPPS for joint planning and resource allocation.[9]

---

[9] GAO-07-883, July 2007: *EPA-State Enforcement Partnership Has Improved but EPA's Oversight Needs Further Enhancement.*

Despite the obstacles and challenges noted in this review, there has been significant progress in advancing NEPPS principles and the EPA-state relationship/partnership. Ongoing trends reveal that NEPPS is on a solid footing but challenges to fully realizing its benefits nevertheless persist. However, there are opportunities for further improvement which are discussed in this review. Although there is some variation in the degree of support among the National Program Managers (NPMs), regions and even the states, overall, leadership support for NEPPS currently is quite strong at both the state and EPA levels (including the regions and NPMs). There is a strong commitment on the part of EPA headquarters and the regions to implement PPGs effectively by providing guidance (e.g., annual NEPPS national guidance; Tribal and State PPG Best Practices Guides). Self-assessments and joint evaluations continue to play an important role in supporting priority setting and documenting accomplishments in meeting goals, objectives, outcomes and outputs.

III. PROGRESS AND CHALLENGES

This section describes observations, challenges, benefits and lessons learned stemming primarily from two key sources that provide a holistic viewpoint of NEPPS: the EPA-funded state planning pilots conducted from 2003 to 2009, and OCIR's interviews of the ten DRAs. Collectively, the results of all these inputs provide historical insight as well as an up-to-date and comprehensive regional-state perspective on how well NEPPS is working.

ECOS selected a number of pilot states and conducted three rounds of pilots between 2003 and 2009 to test new approaches to state and federal planning and priority setting.[10] In particular, the pilots focused on developing state strategic plans and state/regional priorities; internal alignment and improvements in joint planning and priority setting; state engagement in EPA's national and regional planning processes; and developing or improving PPAs/PPGs.[11]

During the fall of 2011 and early 2012, OCIR interviewed the DRAs about the NEPPS process in particular and the EPA-state relationship in general. During the interviews, the DRAs offered observations, described challenges and provided recommendations on a range of topics, including how well NEPPS is implemented in the regions, the use of PPAs/PPGs as tools, the planning process, approaches to improving effectiveness and flexibility, differential oversight, and resource/budget issues.

A. NEPPS Process and Tools

Results of ECOS-EPA Strategic Planning Pilots

Following the first round of pilots, ECOS reported an increase in efforts and opportunities to: better engage the states in joint planning and national strategic planning and priority setting;

---

[10] The following states participated in ECOS pilots, either individually or as part of a multi-state effort (breakdown is by region). Region 1: Massachusetts, Connecticut, Maine, New Hampshire, Rhode Island, Vermont; Region 3: Delaware, Maryland, Pennsylvania, Virginia, West Virginia; Region 4: Kentucky, Tennessee, South Carolina; Region 5: Indiana, Illinois, Minnesota, Wisconsin; Region 6: Texas; Region 8: Colorado, Montana, North Dakota, South Dakota, Utah, Wyoming; Region 10: Idaho, Oregon, Washington.

[11] Reports of the Strategic Planning Pilots, published by ECOS, March 2005 and December 2009.

identify states' individual priorities; overcome the disconnects between strategic and operational planning; work more effectively with EPA to address the most important issues and environmental problems; and share the benefits of the pilots with the regions and other states.

States reported specific benefits in the second round of pilots, including: developing frameworks for improving alignment and integration of planning processes and priorities which were supported at both staff and commissioner levels; increased satisfaction with the identification of state priorities, joint strategies and EPA commitments in PPAs/PPGs or workplans; increased ability to negotiate investments, disinvestments and alternative strategies with EPA; increased communication, trust and cross-program collaboration (internally across state media programs, between states in a region, with EPA regional offices, and/or with external stakeholders).

In round three of the pilots, states reported specific benefits which included improved PPAs and PPGs that promote state/regional priorities due to the direct engagement of regional offices in joint planning (and for some states, senior management's involvement made all the difference in the pilots' successes). Echoing the results in rounds 1 and 2, the states reported that progress had been made in improving relationships between states and EPA, in EPA's planning process, and in enhancing opportunities for EPA/state joint planning. The states made it clear that EPA's continued leadership had created opportunities for more focus on state priorities and state engagement in EPA planning. In at least one pilot, it was reported that engaging interstate media associations helped increase communication and collaboration across programs. The states in several of the pilots actively engaged their staff and/or the public in priority setting or strategic planning.

The lessons learned and observations that emerged from these pilots include:

Sustained Leadership Support Drives Action and Success

- Senior leadership commitment at both the state and EPA levels, e.g., commissioner, Regional Administrator (RA) and DRA, can greatly influence the effectiveness of the PPA/PPG, joint planning and priority setting processes.

- An EPA regional champion is critical to moving a state's interest in alternate compliance monitoring strategies, and states considering pursuing such strategies could benefit from other states' experience in this arena.

- Visible and public affirmation of NEPPS by ECOS and EPA provided the needed support for state and EPA regional efforts to develop PPAs and PPGs.

Strategic Planning/Joint Priority Setting Makes a Difference

- Strategic planning/priority setting can break down internal barriers and help identify cross-media opportunities. However, allegiances to and protection of media programs at EPA and in the states are strong forces that affect efforts to promote strategic thinking, priority setting and alignment of planning processes.

- In addition to planning, renewed focus should be placed on assessing progress and evaluating results.

- States want reduced transaction costs in promoting their priorities or strategies, whether for alternative compliance monitoring or cross-media approaches.

- A clear set of internal priorities helps states in their negotiations with EPA regarding grant workplan commitments, PPAs and PPGs.

- User-friendly and flexible PPAs/PPGs are the most successful in establishing and maintaining long-term participation in NEPPS.

Collaborative Learning Builds Awareness and Reduces Uncertainty

- Tools such as webinars, workshops and ongoing EPA-ECOS workgroups are useful for sharing best practices, successes, lessons learned, and developing actions to improve the EPA-state relationship.

Although the states in these pilots commented on certain aspects of the EPA-state relationship, there were sufficient positive reactions to conclude that NEPPS as originally envisioned is continuing to evolve and improve. It is important to note that over a decade ago, states exercised minimal influence in setting goals, priorities and performance commitments, and NPM guidance was issued by program offices on different schedules making comprehensive planning difficult. Today, by contrast, revamped and retooled processes provide ongoing opportunities for states to influence priorities, strategies, annual performance commitments and budget development. Most states now negotiate PPAs and PPGs as part of their joint planning efforts.

When NEPPS was established in 1995, states had limited flexibility to address alternative state priorities or approaches. Categorical grant funds could be used only for a defined set of activities and it was difficult to fund cross-cutting projects. Now, joint planning and priority setting provide opportunities for states to propose alternative priorities, strategies, and approaches to achieving environmental goals. The completely revised Part 35 grant rule provides for a range of flexibility in how state grant funds can be used.

NEPPS and PPAs/PPGs

The DRAs, during the interviews about NEPPS implementation practices, generally lauded the positive and beneficial aspects of the NEPPS process remarking on how well it works, how efficient, effective and flexible it is, and how ultimately it provides benefits and value to the regions and the states. In particular, some DRAs noted that the process drives more frequent dialogue at all levels which results in marked improvements in communications, performance measures, the sharing of results, and joint strategic planning around significant and meaningful issues. Further, it provides the regions, states, and tribes the ability to adjust PPAs to meet individual (local) needs and requirements, and commit to activities that align with EPA's national priorities.

As for states experiencing budget problems, NEPPS provides the means to identify areas requiring attention and how/where to focus their resources. Another salient observation about the NEPPS process is that it provides a good platform for a discussion on redesigning the role of the regions in relation to oversight, direct execution of programs, and implementation of priorities. Finally, it was noted that the NEPPS process provides a structure and vehicle for a policy-level relationship between EPA and states.

The DRAs also commented on PPAs and PPGs, two popular and widely used NEPPS tools. They noted that PPAs increasingly reflect the true joint priorities of EPA and the states; that PPAs/PPGs are useful tools for rallying around local issues which add credibility and value to them as strategic operating agreements. Broad scale PPAs allow states to provide in-depth assessments or descriptions of their environmental conditions.

The Annual Commitment System (ACS), according to the DRAs, provides a good starting point for PPA/PPG negotiations, and is an effective and transparent process for negotiating state commitments. The DRAs reported that the NPMs are generally receptive to regional/state push-back on specific measures or commitments which can result in a successful re-negotiation. They also stressed the fact that the relationship between state and EPA staff is the most important component of a region's work with its states. Even when the work is difficult and disagreements occur, everyone works to ensure that productive relationships are maintained.

One region is undertaking an evaluation/assessment of its current state of affairs which will include a three-year profile of state staffing; regional staffing and STAG funds; the Priorities and Commitments List[12] (P&C List) and ACS measures; and the status of the region's negotiations with its states. This information will be analyzed in the context of the PPA/PPG process and the results could potentially serve as a best practice approach which can be applied more broadly across the regions.

In terms of challenges to overcome, the DRAs reflected on the issue of differing state and federal fiscal years creating difficulties in aligning state and regional planning processes and getting the states to focus on the regional priorities. Since state planning cycles are different, EPA has found it challenging to accommodate the differences. Also, this longstanding planning alignment issue can cause other funding complications and delays which are further compounded by continuing resolutions.

Finally, it was noted that NEPPS is more difficult to implement in states (e.g., California) that are governed by many different districts headed by decision-making boards that have independent authority. This fragmenting of environmental responsibility makes broad strategic planning and PPG workplan negotiations more difficult.

---

[12] This list is used to communicate overall annual PPA workplan expectations to the region's states. It is customized on a state-by-state basis to reflect what the region believes are the most important elements that each state should focus on in the grant year.

Other PPA and PPG Challenges

NEPPS is first and foremost a system of principles and tools to drive performance, efficiency and flexibility in the EPA-state relationship. However, Performance Partnerships are often equated with PPAs and PPGs, and not with NEPPS' broader strategic vision and goals. Nevertheless, misperceptions about PPGs in some quarters regarding accountability continue to persist. Increasing requirements for more performance measurement and reporting create additional tension between accountability and flexibility.

Further, PPGs are somewhat limited in their leveraging ability because approximately two-thirds of EPA's state grants are not eligible to be incorporated into PPGs. While most of these PPG-ineligible grant funds are those that fund large water and wastewater infrastructure projects not directly linked to state program work, i.e., state revolving funds for Clean Water Act (CWA) and Safe Drinking Water Act (SDWA), there are other ineligible grants that do indeed contribute to state program implementation such as those listed in Appendix A. Also, states receive funds from other federal agencies and/or departments (e.g., U.S. Department of Agriculture, Department of Energy) to accomplish similar environmental goals.

Under 40 CFR Part 35, the Administrator has the authority to add, delete, or change the programs eligible for inclusion in a PPG in guidance or in a regulation. If a new grant program is authorized in the State and Tribal Assistance Grants (STAG) appropriations account, the Administrator can include that new program in the list of PPG-eligible programs.

Even though significant progress has been made towards achievement of NEPPS goals, a few key obstacles and shortcomings remain. The use of NEPPS tools—PPAs and PPGs—remains solid and robust in states and regions that favor them. Despite longstanding and proven track records, PPA usage has leveled off in recent years and some regions have reported not much new interest in PPAs[13] as a tool, but much interest in PPGs.

Expanding PPA/PPG usage, however, by increasing the level of state participation may prove to be somewhat challenging. First impressions can also play an important role in whether or not a state seeks to negotiate a partnership agreement. In addition, early in the history of NEPPS, some states that chose to participate initially decided not to continue participating citing high transaction costs. Recent examples[14] of why states chose not to continue participating include:

- One state was not convinced that PPAs and PPGs provide the desired flexibility.
- The federal share was a small part of the state's budget and transaction costs were too high.
- The early PPAs were too cumbersome and bureaucratic, and once negotiated, they were not used as management tools.
- Changes in political leadership can impact decisions about PPA/PPG use.

---

[13] One region noted that it is difficult to convince its states to engage in discussions on PPAs. Part of the reason appears to be that state directors never saw much value in comprehensive planning, as those plans would become completely overwhelmed by the sheer number of natural disasters that are prevalent in the South: hurricanes, tornadoes, floods, etc.

[14] Source: Performance Partnership Participation and Historical Perspective Table, FY 2012.

- One state was unable to resolve an issue with its region which resulted in non-renewal of its PPA.
- One state's PPA has not been renegotiated since 2007 because, as a strategic document, it did not provide the opportunity to change commitments to address the state's priorities.
- One state believes it can achieve needed flexibility with categorical grants.
- One state used a PPG for the first time in 2005 but requested early termination at the end of FY 2008 because it did not believe that any savings, including administrative, were being realized by combining grants in a PPG. This was due in part to the state decentralizing financial and administrative management functions.

Another challenge involves countering a perception about regional differences in how NEPPS tools are used. ECOS made a specific recommendation (in the context of formal comments on OCIR's past NEPPS national guidance) that "EPA address regional offices' differences in their views and practices relating to PPAs/PPGs." Regional support for PPAs/PPGs, by definition, tends to be varied given the disparate and unique needs of individual states. Although EPA strives for national consistency and a level playing field in its relationship with the states, regional variation in NEPPS implementation practices is prevalent. As NPM for NEPPS, OCIR recognizes these differences, and through guidance, encourages the regions to work with their states to use the NEPPS process and tools in the most effective way possible.

Over time, as NEPPS became increasingly institutionalized within EPA and the states, interest in quantifying its overall benefits grew correspondingly. In 2010, as part of another evaluation, OIG staff developed a preliminary framework and methodology to estimate the monetary benefits of annual strategic planning based on the number of states that use PPAs and PPGs. OCIR is interested in this approach and its potential application to NEPPS in particular. OCIR held meetings with OIG staff to better understand the methodology employed, and plans to further explore the feasibility of developing an estimate of the monetary benefits of implementing NEPPS with the OIG and other EPA program evaluation experts.

B. Performance Measurement and Flexibility

Performance Measures and Reporting

Traditionally, nearly all performance reporting had been based on activity or output measures. However, over a decade ago, with support from the states and EPA using the NEPPS process, Core Performance Measures (CPMs) were developed under the operating principle that reporting measures should make environmental indicators and program outcome measures a priority. CPMs were a limited set of national measures designed to help gauge progress towards protection of the environment and public health, and they were an important step in refining reporting measures.

In 1997, the states and EPA issued a joint statement[15] reaffirming their commitment to use core performance measures as tools to track progress in achieving environmental results. In 2001, the

---

[15] *Joint Statement on Measuring Progress Under the National Environmental Performance Partnership System,* August 14, 1997.

PPG regulations at 40 CFR Part 35 mandated performance reporting and joint evaluation of progress. However, in recent years, the CPM effort has been superseded by EPA's broader implementation of GPRA and now the Government Performance and Results Act Modernization Act of 2010 (GPRAMA). Under GPRA, EPA must set out strategic goals and objectives and measures that will be used to assess progress towards meeting them.

In addition to GPRA, EPA has been working with the states for many years to promote more transparent and consistent reporting of state grants and the results from these grants. For example, EPA also recently issued a grants policy on "State Grant Workplans and Progress Reports," which applies only to the fourteen state categorical grant programs, and requires that workplans and associated progress reports prominently display three Essential Elements: the EPA Strategic Plan Goal; the EPA Strategic Plan Objective; and Workplan Commitments plus time frame. This approach provides regions and states flexibility in negotiating workplans and workplan formats, but at the same time, promotes accountability, Strategic Plan alignment and consistent performance reporting. To further transparency, the policy established the State Grant Information Technology (IT) Application (SGITA) to electronically store workplans and progress reports.

In response to Congressional scrutiny of EPA's grant unobligated balances and grantee unexpended appropriations, as well as state concerns over delays in receiving grant awards, EPA recently issued a grants policy on the "Timely Obligation, Award and Expenditure of EPA Grant Funds." This policy, which is intended to streamline grant processes and improve grant outlay rates, addresses issues relating to delays in obligating grant funds in the first year of availability; delays in awarding grant funds after the passage of a full appropriation; grantee accumulation of unexpended appropriations in awarded grants; and the need to accelerate grant outlays.

Another goal that EPA has been pursuing is reducing states' reporting burdens through actions to increase transparency and direct more attention to the regions' and states' environmental priorities. The EPA-ECOS Reporting Burden Reduction Initiative, which began in 2006 and concluded in 2011, was intended to reduce states' low-value, high-burden reporting requirements for various media (e.g., air, water), thus conserving both states' and EPA's valuable resources while maintaining a commitment to protecting human health and the environment. A substantial number of the priority areas identified by the states have been resolved at the regional and headquarters levels thus resulting in significant reductions in reporting burden, increased transparency, clarity and open communication. This is a significant accomplishment which not only advances NEPPS principles and strengthens the EPA-state relationship in general, but also ensures efficiency by creating tools for incorporating burden reduction into EPA's standard operating procedures.

Today's challenge is to continue to streamline and improve performance measures and reporting, as well as data quality and data systems in order to provide a complete and accurate picture of environmental performance and accomplishments.[16]

Flexibility

From a regional perspective, most of the DRAs interviewed affirmed that PPGs are very helpful; their flexibility highly valued, and in the future will be even more useful as regions and states increasingly face tighter fiscal constraints. States appreciate the additional funding flexibility available through PPGs because it allows them to focus more on higher priorities with multi-year PPGs. PPGs also provide a central point for coordination, especially for meeting match requirements.

Administrative flexibility is the one characteristic most often cited and touted when describing the comprehensive benefits of PPGs. Many DRAs described PPGs as strong, streamlined structures with administrative flexibility and efficiency. The benefits include streamlined accounting and reporting, and the composite cost-share feature eliminates the need to monitor and track the required state match. In addition, states have the ability to pool resources to address priority work (e.g., emergency environmental and public health issues), support special projects, purchase equipment, fund staff, and meet cost-share requirements. See Appendix B for specific examples of flexibility in PPGs.

However, the DRAs noted that the use of programmatic flexibility available through PPGs has been limited. Numerous reasons were offered for the persistence of barriers to greater flexibility in PPGs. Many DRAs noted that flexibility using NEPPS as a vehicle is not occurring because of resource constraints. States and regions are forced to choose from among the statutory mandates by trying to identify and eliminate those causing the least harm and presenting the least risk. Programmatic shifts do not occur within PPGs because of stovepipes and fiscal constraints. In fact, programmatic flexibility is the exception rather than the rule because states rarely ask to shift work between programs. In addition, programmatic flexibility in PPGs is constrained because of the large number of measures and other requirements in NPM guidance, as well as the large number of Program Results Codes (PRCs).[17]

Other roadblocks to PPG flexibility were cited. Sometimes issues that are internal to a state—fee structures, technical or legal constraints, planning and accounting processes, inability to co-mingle funds—can limit its ability to take advantage of available flexibilities. In other instances,

---

[16] EPA began a collaborative effort with the states in 2012 called E-Enterprise. It is a transformative concept intended to make the national environmental system more accessible, effective and efficient by enabling state and federal regulators, regulated entities and the public to take advantage of advances in monitoring, reporting and information technology. For example, information that is now housed in separate databases (or file cabinets) will be electronically integrated and accessible. EPA and the states will have common data to allow collaborative management of the national environmental program.

[17] PRCs were established to account for and relate resources to the Agency's Strategic Plan goals and objectives, national program offices and responsibilities, and governmental functions. PRCs are created when the annual EPA budget is submitted to Congress each February and then formally established in the Integrated Financial Management System (IFMS) with the enactment of EPA's appropriation act. PRCs may be updated at any time.

states would have to change their business practices to make NEPPS work better for them. In certain cases, the region itself becomes the stumbling block. It was observed that some project officers have a difficult time understanding the PPG concept that grant money loses its identity when it becomes part of a PPG. This has been challenging for one region especially with the Brownfields program and managing carryover funds in particular. Also, EPA policy can be an obstacle. For example, the unliquidated obligations (ULO) grant policy may sometimes conflict with PPG flexibility. The policy requires that ULOs be managed within a specified time frame. A state requesting an extension for an expiring project period in a PPG that exceeds the ULO policy's time frame would require a waiver which is typically not granted except under exigent circumstances.[18]

C. State Oversight

The differential oversight concept, originally proposed in the 1995 NEPPS Agreement, has not been implemented. The Agreement envisioned EPA focusing on program-wide, limited, after-the-fact reviews rather than on a case-by-case intervention. The Agreement stated that differential oversight would serve as an incentive for strong state performance, thus enabling EPA to focus resources on state programs that need more assistance to perform well.

Eight years ago, a DRA put forth a proposal which was primarily intended to emphasize environmental results and revitalize the differential oversight component of the 1995 NEPPS Agreement. Its key operational feature involved accreditation of state environmental programs that demonstrated satisfactory performance. The primary vehicle for implementing this proposal would be a 3-year PPA as the single definitive agreement that clearly documented the state's responsibilities to meet federal requirements, including performance measures and expectations. Benefits include the ability of EPA to assist states by greatly reducing the real-time oversight of state permits, inspections, enforcement actions and other state decisions. The proposal was tabled because the author was unable to build sufficient support across the regions to improve and refine his proposal.

Several DRAs acknowledged that differential oversight has never been implemented due to the difficulty in developing appropriate benchmark metrics to assess state/regional performance. Another reason, one DRA noted, was because of the close hold that media programs have on their activities. Tailored or variable oversight is generally avoided in another region because it's too complicated. For example, since state performance varies over time, it would be difficult for a region to determine the level of performance at any given time, and whether/when it changes. In yet another region, differential oversight is implemented on a case-by-case, program-by-program basis. One region emphasized that striking a balance between after-the-fact oversight

---

[18]EPA's ULO policy (GPI-11-01) limits the project period for PPGs to 5 years but does not address situations in which activities are funded during the final year of a 5-year PPG, thus requiring additional time for completion. For example, one state has some older grants with contracts that take longer than 5 years to complete. Seeking additional time requires a waiver, which Senior Resource Officials may request under this policy, but only under limited circumstances—national security concerns, circumstances of unusual or compelling urgency, unique programmatic considerations or the public interest. This also affects "high-risk" grantees since they have a tendency to not complete workplan activities in a timely manner.

(e.g., the State Review Framework (SRF)) and ongoing, real-time reviews (e.g., Permit Quality Reviews (PQR)) appears to be the most optimal approach.

Oversight Reform and EPA Initiatives

While state concerns over redundancies, excessive reporting requirements, unnecessary EPA interventions and uneven oversight, as well as EPA concerns with accountability and leveling the playing field persist, a number of national oversight reforms have been implemented. Some states and regions have even further improved their oversight relationships. EPA has developed several initiatives that lay the groundwork for further oversight reform.

- *The State Review Framework* was instituted to insure consistent oversight of all states for enforcement and compliance assurance regarding the Clean Air Act (CAA), CWA, and the Resource Conservation and Recovery Act (RCRA). The SRF created a more predictable baseline oversight approach and increased transparency of the effectiveness of delegated programs.

- *The CWA Action Plan* reforms oversight to address high noncompliance, low enforcement rates, and absence of data for the National Pollutant Discharge Elimination System (NPDES) program. The Action Plan focuses enforcement on the most serious water pollution problems, strengthens and makes more consistent oversight of states and tribes, and improves public accountability through increased transparency.

- *State Performance Dashboards* and comparative maps provide the public with information about the performance of state and EPA enforcement and compliance programs across the country. The dashboards help ensure a minimum level of state performance and a level playing field for regulated entities across states. Information is currently available for CWA (NPDES), RCRA, and CAA programs. The dashboards will be expanded to include SDWA and the Federal Insecticide, Rodenticide, and Fungicide (FIFRA).

- *The Permitting for Environment Results* (PER) initiative is a multi-year effort by EPA and the states to improve the overall integrity and performance of the NPDES program. Since most states are authorized to implement the NPDES program, the PER initiative is based on a strong partnership between the states and EPA. The program prioritizes permits by environmental significance, identifies best practices for increasing permitting process efficiency, and regularly assesses the effectiveness of NPDES programs. The PER effort assures the consistent implementation of NPDES.

- *Permit Quality Reviews* is a process EPA uses to assess whether NPDES permits are developed in a manner that is consistent with applicable requirements in the CWA and environmental regulations. During each PQR, EPA reviews a representative sample of states' NPDES permits and evaluates permit language, fact sheets, calculations, and any supporting documents in the permit files. Through this review mechanism, EPA can promote national consistency and identify successes in implementation of the NPDES program, as well as opportunities for improvement in the development of NPDES permits.

- *Oversight of State Delegations Key Performance Indicator* (KPI) pertains to state oversight and is an area of concern for the EPA-state partnership despite the progress of national reforms. The OIG first identified state oversight as an Agency Management Challenge in FY 2008. An internal look into this issue revealed that many EPA programs utilize a variety of formal and informal oversight processes that may be inconsistent across regions and states. Also, there is not a clear connection between state oversight and other aspects of the EPA-state relationship, such as third-party petitions to withdraw program delegations and key opportunities to broadly address the relationship through PPAs/PPGs, or other joint operating agreements. Rather than allowing this to become an Agency Level Weakness, the Deputy Administrator has made state oversight an EPA priority and requested a KPI be developed and included in the for the Cross-Cutting Fundamental Strategy for the FY 2012 and FY 2013 Action Plans for Strengthening State, Tribal, and International Partnerships.[19]

Interestingly, one DRA did note that the Office of Enforcement and Compliance Assurance's (OECA) SRF and the CWA Action Plan may be creating a framework for implementing some degree of differential oversight in the future. The SRF could serve as a way to implement differential oversight because it establishes consistent baselines for performance. Although the SRF has the potential to reduce uneven oversight, it increases the need for continued improvement in the completeness and accuracy of data and the efficiency of data systems.

One region put forth an interesting suggestion involving PPAs which could serve as a potential best practice example with broader appeal. In this region, PPAs contain program review protocols wherein the states and the region agree to use the protocols to determine which program reviews to undertake. This allows the region to coordinate scheduling reviews so they do not all occur at once and overwhelm the state.

D. Resource and Budget Issues

The State Budget Crisis

Both EPA and the states fulfill critical roles in protecting and improving human health and the environment. Although the recession ended three years ago, and there are signs of improvement, the recovery has been slow and uneven across the nation. According to the Summer 2012

---

[19] KPI for Oversight of State Delegations: "Through an Agency-wide workgroup (National Program Managers, Regions, and headquarters support offices), plan and implement an Agency-wide effort to collect available information to define, describe, and assess EPA's processes, practices, and tools for overseeing state delegations and authorizations. By September 2013, the workgroup will report its findings to the Deputy Administrator and propose options for next steps as needed to ensure the Agency is carrying out its oversight responsibilities in a coordinated, transparent and accountable manner." *FY 2012 and FY 2013 Action Plans: Strengthening State, Tribal, and International Partnerships.*

National Conference of State Legislatures (NCSL) State Budget Update, state budgets still face considerable challenges. Nonetheless, state budgets today are better positioned to handle these challenges. State year-end closing balances are rising, revenue performance is improving and the states are beginning to return to their pre-recession general fund levels. New budget gaps are rare and confined to a few states and state spending has remained within budgeted amounts. For FY 2013, the prevailing economic outlook is one of continued growth but at a slow to moderate pace in most states. Despite these positive developments, there still remains uncertainty over state budgets with concerns about unemployment levels, federal deficit reduction actions, funding for Medicaid, concern about implementing the federal Affordable Care Act, and other external events (e.g., the European debt crisis) contributing to a cautious state budget outlook.

Since states and their respective state environmental agencies operate the majority of federally delegated programs, having sufficient funds to carry out this responsibility is a significant concern for states. A September 2012 ECOS Green Report, *Status of State Environmental Agency Budgets, 2011-2013*, notes that state environmental agency budgets, despite a modest increase in contributions from state general funds in FY 2012, declined overall in FY 2011 and FY 2012 and budget projections for FY 2013 indicate a similar trend. For FY 2013, most of the states expect increased permit fees and contributions from the state general fund but a reduction in federal funding. States receive most of the funds for their environmental agencies from state-imposed fees, a lesser contribution from the federal government and the smallest contribution from state general funds. The growing significance of federal grant money in state environmental agency budgets, combined with decreasing EPA budgets, strains the ability of states to implement environmental programs. A 2010 EPA White Paper[20] reports that costs of state program activities are rising, exacerbating the budget crisis, and making flexible and effective partnerships more important than ever for accomplishing shared environmental goals.

The issue of regional-state resource and budget constraints generated considerable discussion and concern among the DRAs. States continue to struggle with budgets that in most cases are not expected to improve in the near future. Programmatic demands are increasing while resources are decreasing thus making tradeoffs among core program activities difficult. Some regions are concerned that some states may actually begin to give programs back to EPA. With smaller budgets, state agencies cannot accomplish the same amount of work, and it is becoming more difficult to meet present and future EPA national commitments/targets. Yet, despite this, many states continue to commit to EPA's national commitments/targets in their workplans. On the regional front, there is some concern that if there are staff cutbacks due to possible EPA budget reductions, the regions are less likely to absorb work that the states are unable to accomplish.

Commendably, regions have tried to help states address resource issues through a variety of worksharing approaches: one region is engaging effectively in worksharing through PPAs/PPGs (e.g., since EPA has primacy for clean water implementation in two states, the region coordinates extensively with both to avoid duplication of effort). Another region has engaged in

---

[20] *White Paper on Meeting the Challenges of Environmental Protection Together: Building Strong EPA-State Partnerships*, EPA, October 2010.

worksharing with its states on an informal basis by undertaking certain commitments to meet national targets (e.g., if 30 permits are required of a state, the state might do 20, while the region agrees to complete the other 10). A third region has entered into a worksharing agreement with one of its states to do underground storage tank (UST) inspections because of state budget shortfalls. A fourth region has used intergovernmental personnel assignments (IPAs) to bring state staff to EPA in order to keep them working and maintain their expertise while the state was undergoing a reduction in force (RIF).

One DRA described a higher level management hurdle that needs to be overcome. It involves a culture where the region's staff and managers attempt to do everything possible to meet commitments despite limited resources. Although laudable as a goal, it is not a sustainable long-term path and the problem will become exacerbated in the coming years due to state and EPA resource constraints. Another DRA felt that improving final funding allocations is a key issue that should be improved. More often than not, due to continuing resolutions, the region doesn't receive its final funding allocation until well into a state's fiscal year (and many states have differing fiscal years). Until that happens, the region must work with a draft allocation, which of course differs from the final version, and realigning the two when the funding allocation becomes final is time consuming and labor intensive.

Recent Developments in Worksharing

The principles and processes embodied in the original NEPPS Agreement are designed to foster flexible and effective worksharing arrangements between EPA and the states. Budget constraints caused by the economic downturn and slow recovery have generated more interest and attention on worksharing as a tool to help states and EPA implement programs more efficiently.[21]

The EPA-State Worksharing Task Force which was formed in 2011 and published a report in 2012 on worksharing prohibitions and areas of caution.[22] The prohibitions identified were selection of a Superfund site remedy and preparation of a state's competitive grant application. The areas of caution primarily dealt with grants and appropriations law. The Task Force published a second report on principles and best practices for worksharing in March 2013.[23] It provides a set of principles and best practices for several commonly employed worksharing scenarios, plus two regional examples. The Task Force expects to publish another report in calendar year 2013 addressing training opportunities for state staff.

---

[21] EPA's recent commitment to worksharing is evidenced in the FY 2011– 2015 EPA Strategic Plan, which includes a Cross-Cutting Fundamental Strategy that incorporates worksharing. The 2013 Action Plan, which implements the cross-cutting strategy, includes a Key Performance Indicator for regions to develop implementation targets for worksharing. See: http://epa.gov/planandbudget/annualplan/Strategy_4_FY_2013_Action_Plan.pdf.

[22] See: http://www.epa.gov/ocir/nepps/pdf/task_force_report_prohibitions_areas_caution.pdf.

[23] See: http://www.epa.gov/ocir/nepps/pdf/task_force_reportbstpractices.3.26.13.pdf.

The Task Force found:

- NPM support for worksharing efforts is critical to regions seeking to expand worksharing opportunities with states;
- NPMs should strengthen their support for worksharing in their annual guidance;
- If worksharing is negotiated as part of the annual planning process between a region and state, a PPA or PPG can be used as a tool to document the results of the negotiation;
- Worksharing agreements also can be negotiated on an as-needed basis to address unique or unplanned situations that may arise in a state, such as a natural disaster.

In terms of challenges, future constraints on EPA's budget may hinder the regions' ability to take on additional work through worksharing arrangements. The Task Force also noted that while PPAs, PPGs, or other grant workplans provide one means for documenting worksharing, currently there is no Agency-wide system to track implementation.

IV. RECOMMENDATIONS

The following recommendations stem from the various inputs in this review, mostly from the DRA interviews and a few from the EPA-state planning pilots and other third-party evaluations, and are grouped under these categories: the NEPPS process and tools; performance measurement and flexibility; state oversight; resource and workload issues.

Overall, NEPPS is viewed as a valuable and beneficial process for advancing and strengthening the EPA-state partnership and deemed worthy of continued federal/state revitalization efforts. Senior leadership commitment (state and regional) is essential to break down internal barriers, help to identify cross-programmatic opportunities, and improve strategic planning/priority setting. However, it was also noted that allegiances to and protection of media programs at EPA and in the states are strong forces that make broadening the use and scope of NEPPS difficult.

NEPPS Process and Tools

- Pilot projects between EPA and states to improve joint priority setting and strategic planning have been effective in advancing Performance Partnerships. Consider implementing a new series of EPA-state planning pilots to initiate new Performance Partnerships or strengthen existing ones.

- To revitalize and reemphasize the value and benefits of NEPPS:

  o Provide more visible senior leadership support and commitment among the states and within EPA to: publicly and visibly affirm NEPPS; promote and advocate for the use of PPAs/PPGs as tools for implementing environmental programs; and provide the necessary support for state/regional efforts to use the NEPPS approach to organize the EPA-state relationship.

  o Increase the use of training and outreach to the states, the regions and headquarters (e.g., via webinars, workshops, conference calls, PPG training modules, joint training on best practices and the reconciliation of tensions between flexibility and accountability).

- Develop an approach to quantify the joint planning benefits of NEPPS.

- Build stronger ties between NEPPS and the tribes to ensure best practices and consistent implementation of EPA policies and regulations.

- Enhance the effectiveness of PPAs and PPGs:

  o By seeking to achieve better integration of the Administrator's priorities (e.g., EJ, children's health) into core program work.

  o Through the broader application of EPA's policy on in-kind assistance to make EPA contract vehicles available to the states via PPGs or categorical grants.[24]

  o By distinguishing clearly in PPAs between activities funded by EPA (which should be considered grant commitments) and those that are state funded.

- Investigate opportunities to work with other government agencies and/or departments to identify potential opportunities to apply the PPG model to interagency pooled funding for similar environmental purposes.[25]

---

[24] Regarding "in-kind assistance," EPA Order 5700.1 (3/22/94) states that in addition to transferring money, EPA may use assistance agreements to transfer anything of value, such as equipment or services, to an authorized assistance recipient. If it would be more efficient in terms of cost or time for EPA rather than the recipient to purchase equipment or services, EPA may do so and provide the equipment or services to the recipient under the assistance agreement. This approach would be appropriate, for example, where a piece of equipment necessary for a grant-assisted project can be purchased at a considerably lower price or be delivered much earlier using an existing EPA contract. Likewise, EPA may provide the services of an EPA contractor to an assistance recipient in lieu of money, where appropriate.

[25] During the summer of 2011, OCIR participated in an OMB workgroup formed to implement the Presidential Memorandum on "Administrative Flexibility, Lower Costs, and Better Results for State, Local, and Tribal Governments." The workgroup included 10 states, 10 federal agencies, and several associations. Its charge was to seek out opportunities to improve financial management of federal funds, with a specific emphasis on ways to focus more on performance and outcomes rather than redundant and overly prescriptive reporting on how funds are spent. As part of the workgroup's deliberations, OCIR provided several briefings on NEPPS and PPGs which were well received and generated a great deal of interest and discussion among workgroup members. The final report, which was issued in the fall of 2011, included recommendations to develop pilot programs that would allow states to blend federal funds from similar programs within the same agency or even multiple agencies. The PPG model was referenced and the administrative benefits and flexibility of PPGs were cited as goals of the pilots. The appropriations bill for Labor, HHS and Education (S. 3295), which passed the Senate Appropriations Committee, allocated a maximum of $130,000,000 of discretionary funds to fund up to 13 Performance Partnership Pilots provided that such pilots are targeted at programs serving disconnected youth. The bill defined a Performance Partnership Pilot as a project designed to identify cost-effective strategies for providing services at the state, regional, or local level. The Committee Report accompanying S. 3295 described Performance Partnerships as "a new authority that will provide States and local communities with unprecedented flexibility to achieve defined outcomes."

- Update the PPG regulations at 40 CFR Part 35 to include all categorical grants that are PPG eligible.[26]

Performance Measurement and Flexibility

- Value flexibility, especially in accountability systems by reducing the number of ACS measures and PRCs, reducing reporting burden, and increasing the ability to focus on regional and state needs.

- Continue to ensure the timely consideration of and decisions on state requests for flexibility and innovation, and ensure that EPA regions support proposals for alternate compliance monitoring strategies.[27]

State Oversight

- Develop a consistent approach that specifies the regional and state roles and responsibilities for implementing the SRF or any other program review recommendations during the PPA/PPG negotiation process.

- Allow for more flexibility to better align state environmental priorities with national targets (e.g., by shifting emphasis from major to minor sources since the latter is of greater concern to some states).

- Use Information Technology to improve accountability and streamline business processes, from workplans to reporting, and automate the process to replace annual or semi-annual visits/audits, particularly for states that perform well. This approach could then serve as a kind of differential oversight.

Resource and Workload Issues

- NPMs should integrate into the strategic planning process a more explicit and systematic approach for making investments and disinvestments, which will better enable the regions and the states to address top priorities while maintaining core programs activities.

- Analyze past state performance to develop a baseline of state performance data to inform the EPA budget and NPM ACS targets. Currently, there is little analysis or consideration of state contributions in developing EPA's commitments in the Annual Performance Plan.

---

[26]Appendix A lists all the grants that are not currently but can be made eligible for inclusion in PPGs. However, only one of the grants listed in the Appendix is currently funded.

[27] In February 2013, EPA approved Massachusetts' 3-year alternative compliance monitoring strategy for federally-funded inspections for the air, RCRA and underground storage tank programs. EPA's approval authorizes the state to implement a more effective compliance assurance program that tailors EPA's national compliance strategy to fit actual conditions in the state and maximize limited resources.

- Discuss worksharing when NPM guidance is being formulated. More explicit NPM guidance on what work could/should be shared to achieve national or regional targets will help in negotiating with the states and with internal regional work planning.

- Ensure that worksharing is part of the annual planning discussions between regions and states. Document worksharing agreements in PPAs, PPGs, or other grant workplans, and develop a means to track those that occur outside of annual planning discussions.

- Implement the recommendations of the EPA-State Worksharing Task Force and consult its reports for guidance in developing worksharing agreements.

- Develop an approach to identify the scope and costs of core program work conducted by EPA and the states in order to help regions understand the variation in state implementation costs.

V. REFERENCES

ECOS Resolution Number 8-10. Revised March 30, 2011. *Continued State Commitment to NEPPS and Strengthening the State-EPA Partnership.*

ECOS Website (www.ecos.org). November 2010. *Summary Data: Delegation by Environmental Act.*

ECOS-EPA Reports. March 2005 and December 2009. Strategic Planning Pilots Reports.

Elkins, Arthur A. Jr., March 2, 2011. *Major Management Challenges at the Environmental Protection Agency. Testimony of Arthur A. Elkins, Jr., EPA Inspector General, before the Subcommittee on Interior, Environment, and Related Agencies Committee on Appropriations, U.S. House of Representatives.*

GAO-10165T, October 15, 2009: *Clean Water Act: Longstanding Issues Impact EPA's and States' Enforcement Efforts.*

GAO-07-883, July 2007: *EPA-State Enforcement Partnership Has Improved, but EPA's Oversight Needs Further Enhancement.*

GAO-06-840T, June 2006: *EPA's Effort to Improve and Make More Consistent Its Compliance and Enforcement Activities.*

GAO-06-625, May 2006: *EPA has Made Progress in Grant Reforms but Needs to Address Weaknesses in Implementation and Accountability.*

GAO-05-52, November 2004: *Better Coordination is Needed to Develop Environmental Indicator Sets that Inform Decisions.*

GAO-03-112, January 2003: *Major Management Challenges and Program Risks: Environmental Protection Agency.*

GAO-03-846, August 2003: *EPA Needs to Strengthen Efforts to Address Persistent Challenges.*

GAO-01-257, January 2001: *Major Management Challenges and Program Risks: Environmental Protection Agency*

GAO-01-774, June 2001: *Environmental Protection Agency: Status of Achieving Key Outcomes and Addressing Major Management Challenges.*

GAO/RCED-00-108, June 2000: *More Consistency Needed Among EPA Regions in Approach to Enforcement.*

GAO/RCED-99-171, June, 1999: *Collaborative EPA-State Effort Needed to Improve New Performance Partnership System.*

GAO/RCED-98-113, May 1998: *EPA's and States' Efforts to Focus State Enforcement Programs on Results.*

GAO/T-RCED-96-87, February 1996: *Status of EPA's Initiatives to Create a New Partnership with States.*

GAO/RCED-96-41, February 1996: *An Integrated Approach Could Reduce Pollution and Increase Regulatory Efficiency.*

Guerrero, Peter F. May 2, 2000. GAO/T-RCED-00-163: *Collaborative EPA-State Effort Needed to Improve Performance Partnership System. Testimony of Peter F. Guerrero, Director, GAO, before the Committee on Environment and Public Works, United States Senate.*

Mittal, Anu K. October 15, 2009. GAO-09-165T: *Clean Water Act: Longstanding Issues Impact EPA's and States' Enforcement Efforts. Testimony of Anu K. Mittal, Director, GAO, before the Committee on Transportation and Infrastructure, U.S. House of Representatives.*

NAPA Report. April 2007. *Taking Environmental Protection to the Next Level: An Assessment of the U.S. Environmental Services Delivery System.*

NAPA Report. November 2000. *Transforming Environmental Protection for the 21st Century.*

OIG Report No. 12-P-0113, December 2011: *EPA Must Improve Oversight of State Enforcement.*

OIG Report No. 2000-P-00025, December 2011: *North Carolina NPDES Enforcement and EPA Region 4 Oversight.*

OIG Report No. 11-P-0315, July 2011: *Agency-Wide Application of Region 7 NPDES Program Process Improvements Could Increase EPA Efficiency.*

OIG Report No. 10-P-0224, September 2010: *EPA Should Revise Outdated or Inconsistent EPA-State Clean Water Act Memoranda of Agreement.*

OIG Report No. 08-P-0278, September 2008: *EPA Has Initiated Strategic Planning for Priority Enforcement Areas, but Key Elements Still Needed.*

OIG Report No. 2006P-00006, December 2005: *EPA Performance Measures Do Not Effectively Track Compliance Outcomes.*

OIG Report No. 2003-P-00004, December 2002: *EPA Needs to More Actively Promote State Self Assessment of Environmental Programs.*

OIG Report 2000-M-000828-000011, March 2000: *EPA Needs Better Integration of the National Environmental Performance Partnership System.*

OIG Report No. 2000-P-00008, February 2000: *Improving Region 5's EnPPA/PPG Program.*

OIG Report No. 1999-000209-R8-100302, September 1999: *Region 8 Needs to Improve its Performance Partnership Program to Ensure Accountability and Improved Environmental Results.*

OIG Report No. 1999-P00216, September 1999: *Region 4's Implementation and Oversight of Performance Partnership Grants.*

OIG Report No. 1999-000208 –R6-100282, September 1999: *Region 6 Oversight of Performance Partnership Grants.*

Ross and Associates Report. November 1999. *How Well is NEPPS Working? A Summary Comparison of Several Recent Evaluations of the National Environmental Performance Partnership System.*

Stephenson, John B. March 3, 2004. GAO-04-510T: *EPA Needs to Strengthen Efforts to Address Management Challenges. Testimony of John B. Stephenson, Director, GAO, before the Committee on Environment and Public Works, United States Senate.*

Stephenson, John B. July 20, 2004. GAO-04-983T: *EPA Continues to Have Problems Linking Grants to Environmental Results. Testimony of John B. Stephenson, Director, GAO, before the Subcommittee on Water Resources and Environment, Committee on Transportation and Infrastructure, U.S. House of Representatives.*

Stephenson, John B. October 1, 2003. GAO-04-122T: *EPA Needs to Strengthen Oversight and Enhance Accountability to Address Persistent Challenges. Testimony of John B. Stephenson, Director, GAO, before the before the Subcommittee on Water Resources and Environment, Committee on Transportation and Infrastructure, U.S. House of Representatives.*

Stephenson, John B. June 11, 2003. GAO-03-628T: *Environmental Protection Agency: Problems Persist in Effectively Managing Grants. Testimony of John B. Stephenson, Director, GAO, before the Subcommittee on Water Resources and Environment, Committee on Transportation and Infrastructure, U.S. House of Representatives.*

Tinsley, Nikki. 2002. GAO-2001-P-00010: EPA's Top 10 Management Challenges. *Testimony of Nikki Tinsley, Inspector General, USEPA, before the Subcommittee on Energy Policy,*

Natural Resources and Regulatory Affairs, Committee on Government Reform, U.S. House of Representatives.

U.S. EPA/OCIR. 2005-2013. *NEPPS National Guidance.*

U.S. EPA. *Fiscal Years 2011-2015 EPA Strategic Plan Cross-cutting Fundamental Strategy for Strengthening State, Tribal and International Partnerships.*

U.S. EPA. July 2012. Report of the EPA-State Worksharing Task Force: *Prohibitions, Areas of Caution and Recommendations to Enhance Worksharing Opportunities.*

U.S. EPA. March 2013. Report of the EPA-State Worksharing Task Force: *Principles and best Practices for Worksharing.*

U.S. EPA/OCIR. Fall 2011 and Winter 2012. Interviews of the Deputy Regional Administrators on NEPPS and the EPA-State Partnership.

U.S. EPA. October 1, 2010. *Grants Policy Issuance 11-01: Managing Unliquidated Obligations and Ensuring Progress under EPA Assistance Agreements.*

U.S. EPA. October 1, 2012. *Grants Policy Issuance 11-03: State Grant Workplans and Progress Reports.*

U.S. EPA. October 2010. *White Paper on Meeting the Challenges of Environmental Protection Together: Building Strong EPA-State Partnerships.*

U.S. EPA/OCIR. *2010 NEPPS Program Implementation Summary.*

U.S. EPA/OCIR. *2010 PPA-PPG Current Status–Historical Perspective Table.*

U.S. EPA-State Alignment and Performance Partnership Agreement Workgroup Report. January 2005. *Overview of Alignment and Performance Partnership Process Improvements.*

U.S. EPA-State Joint Statement. August 14, 1997. *Joint Statement on Measuring Progress Under the National Environmental Performance Partnership System.*

U.S. EPA-State NEPPS Agreement. May 17, 1995. *Joint Commitment to Reform Oversight and Create a National Environmental Performance Partnership System.*

U.S. EPA. March 22, 1994. EPA Order 5700.1: *Policy for Distinguishing Between Assistance and Acquisition.*

U.S. EPA. May 17, 1993. Task Force to Enhance State Capacity Report: *Strengthening Environmental Management in the United States.*

PPG Eligible Grants Not Included in Original STAG Appropriations Earmarks			
Grant	Purpose	PPG Eligible?	Funding Status
Targeted Watersheds	Provides resources to support watershed organizations in their efforts to expand and improve existing water protection measures.	Yes, but not yet added to the list of programs authorized to be included in PPGs.	$0.0 Last funded in FY 2011
Homeland Security (Water Security Initiative)	Provides resources to assist with critical water infrastructure protection (e.g., through technical assistance, training, and communication activities).	Yes, but not yet added to the list of programs authorized to be included in PPGs.	$0.0 (FY 2012 Enacted Budget and FY 2013 PB) Last funded in FY 2009 when program was in STAG; now funded under Science and Technology.
Beaches Protection	Funds the 35 states and territories with Great Lakes or coastal shorelines for protecting public health at the nation's beaches.	Yes, but not yet added to the list of programs authorized to be included in PPGs.	$9,864,000 (in FY 2012 Enacted Budget) $0.0 (FY 2013 PB) Delta: ($9,864,000)
Wastewater Operator Training (CWA sec. 104(b)(3))	Training of operators at wastewater treatment facilities.	Yes, but not yet added to the list of programs authorized to be included in PPGs.	$0.0 Grant ended in FY 2007
Underground Storage Tanks (UST) (SWDA sec. 2007(f)(2)); LUST Trust Fund	To assist states, territories and tribes in the development and implementation of approved underground storage tank release detection, prevention and correction programs.	No, but EPA policy for the LUST grant program could be clarified legislatively. LUST funds are not eligible for inclusion in PPGs because of a mandate in the Energy Policy Act of 2005 which authorized the use of appropriations from the LUST Trust Fund for assistance agreements to states for leak detection, prevention and related enforcement. Prevention activities are funded under a separate statutory line item within the LUST appropriation. The LUST prevention appropriation is available for a different purpose from the statutory line item for LUST corrective action and the funds are not interchangeable. UST funds are PPG eligible under the authority of SWDA sec. 2007(f)(2) for basic programmatic functions not otherwise authorized under LUST program funding and provided to the states through STAG funding.	$1,584,000 (FY 2102 Enacted Budget) $1,490,000 (FY 2013 PB) Delta: ($58,000) UST funding was $11-12 million pre-Energy Policy Act; post statute funding was $2.5 million annually through FY 2011

Examples of Flexibility in PPGs

Benefit	Example
Address emergency situations and changing conditions	**AL:** Used a multi-year PPG to divert manpower to the BP oil spill effort realizing that if certain grant commitments were not achieved due to the shift in manpower, the state could tap into the extra time frame built into the PPG. **MS:** Used PPG flexibility to deal with issues stemming from the 2011 tornado and flooding disasters. The ability to move funds immediately for response and recovery work in the affected areas was very helpful to the state. **IA:** In August 2008, parts of Iowa experienced flooding from substantial rainfalls. EPA Region 7 awarded additional funds in the state's PPG to enable it to complete water monitoring in twenty-five targeted areas where the flooding was most severe. The funds awarded were PPG carryover funds reprogrammed from another state's closed out PPG. **IL:** The PPG provides the state the ability to pool resources to address priority work. For example, CWA Section 106 and 319 funds are pooled to develop TMDLs that address both point and nonpoint sources of pollution. In addition, the PPG allows the use of pooled resources to address emergency environmental and public health issues such as those caused by the 2011 flooding.
Address state-identified priority/support special project	**GA:** Used its PPG to accomplish priority work on a specific program (water). Since air, water (both 106 and drinking water) and RCRA funding were in the PPG, GA was able to combine small savings from each program to fund water quality studies, water flow studies, and additional monitoring to better document available drinking water sources and assess the potential weather impacts on them. **GA:** Also assisted the metropolitan public water system suppliers in developing watershed management plans to protect their drinking water supplies using funds deobligated from previous year grants. GA would not have been able to accomplish this without a PPG because the deobligated funds from the water grant programs alone would not have been sufficient to fund these additional efforts. Also, EPA funding for the water grant programs was not sufficient to cover the state's needs. **TX:** Used its PPG to fund a special project involving NPDES permitting and enforcement in all water pollution control programs by continuing the development of the Permitting and Registration Information System (PARIS) database project. The project strengthened planning efforts for implementing the CWA Action Plan and integrated data

Benefit	Example
	reporting. The project benefitted the state by providing for business process and systems analysis documentation and improved the state's ability to identify, collect and provide timely, accurate and complete data for reporting to EPA's ICIS-NPDES database. The project was funded from savings realized from each of the programs in the PPG and then awarded into the next year's grant. **CO:** Used PPG flexibility to continue its long-standing emphasis on Pollution Prevention (P2) as the pollution control tool of choice and the incorporation of pollution prevention into state regulations, compliance assistance, enforcement and permitting activities. The P2 program received supplemental funding from each program which has integrated pollution prevention in its core work. The supplemental funding provided staffing support, technical assistance to recipients of both recycling and advanced technology grant programs, and greater public outreach.
Meet cost share requirements	**WA:** Uses extra water match funds to help meet the CAA 105 maintenance of effort (MOE) contribution in its PPG. **UT:** Uses PPG flexibility to ensure that the state can provide sufficient match for all the programs in the PPG. For various timing reasons regarding when federal funds are spent and the availability of state funds for match, the state uses available matching funds from one program to meet the needs of another.
Redirection of carryover funds to purchase equipment, fund staff	**WA:** In situations where there are carryover funds from a closed grant, the state typically redirects them to another program in the PPG (e.g., a water project). **UT:** Redirected PPG funds, mainly from section 105, to fund an FTE in Region 8's Office of Planning and Public Affairs to work on public outreach and involvement in the SIP for air quality. In the past, carryover funds were used to finance partnership efforts in the Unita Basin and southwestern parts of the state. Carryover funds also have been used purchase lab equipment for testing samples from various mediums. **ND:** Purchased new lab equipment to replace existing equipment using carryover funds.
Address state priority by shifting work from a lower to higher priority program area	**ME:** With regional support, shifted resources from TMDL development to TMDL implementation which gave it the ability to reissue all the priority NPDES permits in the Androscoggin River Basin with water-quality based permits for nutrients and biochemical oxygen demand (BOD). This allowed the state to focus its resources to remediate a long-standing water quality problem. **VT:** Shifted staff from lower priority programs to higher priority ones thus allowing the state to efficiently utilize

Benefit	Example
	dwindling federal and state funding in the most effective manner possible to obtain results.
	VI: Would use its PPG to focus on higher priority programs such as drinking water (PWSS program) or water quality (CWA 106) in the event they are impacted by storms during the hurricane season, and still comply with the rest of PPG workplan commitments at the end of the project period. Also, PPG flexibility makes it easier for VI to combine projects/initiatives that deal with air pollution monitoring and the effect of air pollution on the quality of water in cisterns which is a critical water source for its citizens.
	NJ: Revenues generated from environmental fees and fines are reserved for specific purposes, and expenditures are limited by the amount of revenue realized. Under its PPG, NJ deposited many of these dedicated monies into the state's General Fund. Such a shift allows NJ greater flexibility in allocating resources to high priority environmental issues.
Fund cross-cutting projects/initiatives	**CO:** Funds a number of cross-cutting projects and initiatives–one integrates air, water and waste inspections and compliance assistance for animal feeding operations. Others deal with permit and environmental impact reviews; outreach to federal, state and local authorities dealing with the state's rapidly expanding energy industries; information management; the Environmental Leadership Program; and the Pollution Prevention program.
	MO: In FY 2006 and 2008, the state requested flexibility to use PPG-eligible funds for a cross-media permit initiative. As part of the permitting process, MO followed up on each newly-issued permit, environmental concern received from a citizen or other source, or facilities never before inspected with an Environmental Assistance Visit (EAV). The purpose of the EAV was to: 1) ensure that the responsible parties understood the permit requirements; 2) verify the conditions of the permit were being met; 3) investigate any concerns with the permittee or other operation; 4) provide assistance to help achieve compliance where needed; and 5) follow up to ensure environmental performance is satisfactory. These EAVs were conducted for permits eligible under the water, air, and RCRA programs funded in the PPG.
	AZ: The PPG eases the administrative transactions and costs for the state and EPA when funding cross-cutting water projects and initiatives since the state's PPG includes only water grants.
	NJ: Uses PPG funds to provide current information on the state's environmental conditions by maintaining and updating its *Environmental Trends Report*. There are forty-eight chapters and each chapter describes a specific area in

Benefit	Example
	which the state has been working to improve conditions, and presents a specific environmental measure or category of measurements meaningful in gauging the current status of the environment in NJ. The *Environmental Trends Report* includes chapters that address cross-cutting issues: Climate Change, Greenhouse Gas Emissions, Energy Use, Mercury Emission, and Pollution Prevention. There was an upgrade to the state's data systems using multiple program funds through the PPG. Working closely with Region 2's Information Systems Branch, discretionary funds were added to the PPG for this project.
Reduce administrative burden, provide financial flexibility	**MN:** Used PPG flexibility to improve the flow of funds during periods of unpredictability, such as changing budget amounts from year to year within some programs (sometimes with very late notice even during the year); changes in timing of receipt of funds due to continuing resolutions and to differences in timing from program to program; addressing seasonally-related cash-flow challenges from program to program within the PPG generally. During this time of diminishing resources, MN appreciates the administrative burden reduction aspect of PPGs (especially consolidated and simplified reporting). **CT:** With state environmental budgets being reduced, the flexibility provided by the PPA/PPG structure lowers transaction costs and allows the state to use the smaller amount of federal funding in the most effective manner possible. **IL:** Benefits greatly from the administrative efficiencies of streamlined accounting and reporting provided by PPGs, and the composite cost share feature eliminates the need to constantly monitor and track match resources by specific grant.

www.ingramcontent.com/pod-product-compliance
Lightning Source LLC
Chambersburg PA
CBHW081410170526
45166CB00010B/3287